M336
Mathematics and Computing: a third-level course

GROUPS & GEOMETRY

UNIT GR1
PROPERTIES OF THE INTEGERS

Prepared for the course team by
Bob Coates & Bob Margolis

The Open University

This text forms part of an Open University third-level course.
The main printed materials for this course are as follows.

Block 1
Unit IB1   Tilings
Unit IB2   Groups: properties and examples
Unit IB3   Frieze patterns
Unit IB4   Groups: axioms and their consequences

Block 2
Unit GR1   Properties of the integers
Unit GR2   Abelian and cyclic groups
Unit GE1   Counting with groups
Unit GE2   Periodic and transitive tilings

Block 3
Unit GR3   Decomposition of Abelian groups
Unit GR4   Finite groups 1
Unit GE3   Two-dimensional lattices
Unit GE4   Wallpaper patterns

Block 4
Unit GR5   Sylow's theorems
Unit GR6   Finite groups 2
Unit GE5   Groups and solids in three dimensions
Unit GE6   Three-dimensional lattices and polyhedra

The course was produced by the following team:

Andrew Adamyk (BBC Producer)
David Asche (Author, Software and Video)
Jenny Chalmers (Publishing Editor)
Bob Coates (Author)
Sarah Crompton (Graphic Designer)
David Crowe (Author and Video)
Margaret Crowe (Course Manager)
Alison George (Graphic Artist)
Derek Goldrei (Groups Exercises and Assessment)
Fred Holroyd (Chair, Author, Video and Academic Editor)
Jack Koumi (BBC Producer)
Tim Lister (Geometry Exercises and Assessment)
Roger Lowry (Publishing Editor)
Bob Margolis (Author)
Roy Nelson (Author and Video)
Joe Rooney (Author and Video)
Peter Strain-Clark (Author and Video)
Pip Surgey (BBC Producer)

With valuable assistance from:

Maths Faculty Course Materials Production Unit
Christine Bestavachvili (Video Presenter)
Ian Brodie (Reader)
Andrew Brown (Reader)
Judith Daniels (Video Presenter)
Kathleen Gilmartin (Video Presenter)
Liz Scott (Reader)
Heidi Wilson (Reader)
Robin Wilson (Reader)

The external assessor was:

Norman Biggs (Professor of Mathematics, LSE)

The Open University, Walton Hall, Milton Keynes, MK7 6AA.

First published 1994. Reprinted 2002, 2007.

Copyright © 1994 The Open University

All rights reserved. No part of this publication may be reproduced, stored in a retrieval system or transmitted in any form or by any means, without written permission from the publisher or a licence from the Copyright Licensing Agency Limited. Details of such licences (for reprographic reproduction) may be obtained from the Copyright Licensing Agency Ltd of 90 Tottenham Court Road, London, W1P 9HE.

Edited, designed and typeset by the Open University using the Open University TeX System.

Printed in Malta by Gutenberg Press Limited.

ISBN 0 7492 2163 1

This text forms part of an Open University Third Level Course. If you would like a copy of *Studying with the Open University*, please write to the Central Enquiry Service, PO Box 200, The Open University, Walton Hall, Milton Keynes, MK7 6YZ. If you have not already enrolled on the Course and would like to buy this or other Open University material, please write to Open University Educational Enterprises Ltd, 12 Cofferidge Close, Stony Stratford, Milton Keynes, MK11 1BY, United Kingdom.

CONTENTS

Study guide	4
Introduction	5
1 Division	5
2 Factors and multiples	12
3 The Euclidean algorithm (audio-tape section)	17
4 Primes	22
5 Unique prime factorization	26
Solutions to the exercises	29
Objectives	39
Index	40

STUDY GUIDE

You may well be familiar with many of the results in this unit, although perhaps not with this level of formality. As a result, you should not find the study time overly long.

One objective of the unit is that you should be able to use the computational methods and results discussed. It is also important that you are able to follow the proofs and the strategies used, since another objective of the unit is that you should be able to construct similar proofs.

The five sections probably require roughly equal study times.

There is an audio programme associated with Section 3 of this unit.

There is no video programme associated with this unit or with any of the other (GR) units in the Groups stream.

You will not need the *Geometry Envelope* in your study of this unit, nor in any of the other units in the Groups stream.

INTRODUCTION

In this unit we shall be concerned with some of the basic properties of the integers, $\mathbb{Z}$.

We have already used some of the basic properties of $\mathbb{Z}$ in *Units IB2* and *IB4*. We shall continue to need to use such basic properties of $\mathbb{Z}$ in the rest of the Groups stream of this course. It is going to be useful, therefore, in this unit, to gather together the most important results about $\mathbb{Z}$ that we shall need.

Properties of the integers are also used in some of the Geometry units.

In Section 1 we discuss *division* of one integer by another; not in the sense of exact divisibility, but in terms of quotients and remainders.

Section 2 is concerned with properties of 'exact' divisibility, particularly *highest common factors* (HCFs) and *lowest common multiples* (LCMs).

Section 3 deals with the *Euclidean Algorithm*, which gives a practical method of calculating HCFs *without* resorting to the method of factorizing into primes.

Mention of *primes* takes us to the material in Section 4, which includes the definition of primes and the possibility of factorization into primes.

Finally, in Section 5, we show that the decomposition into primes is unique.

1 DIVISION

In general, division is not possible in $\mathbb{Z}$. For example, both 22 and 6 are in $\mathbb{Z}$ but the quotient

$$\frac{22}{6} = \frac{11}{3}$$

is not in $\mathbb{Z}$. Thus division is *not* a binary operation for $\mathbb{Z}$.

The set $\mathbb{Z}$ does not contain fractions.

However, we can still talk about division in $\mathbb{Z}$, in the sense of finding both a *quotient* and *remainder*. For example, we can divide 22 by 6 to obtain a quotient of 3 and a remainder of 4. We can write these facts entirely in $\mathbb{Z}$ as follows:

$$22 = 6 \times 3 + 4.$$

Shortly we shall prove that we can always do this form of division, that is we can write

number = divisor × quotient + remainder.

Actually, we can prove a little more: that the remainder can always be chosen to be non-negative and less than the modulus of the divisor. Furthermore, with these conditions on the remainder, both quotient *and* remainder are unique.

In the special case where the remainder is zero, we say that one number *divides* the other or, equivalently, that one number is a *factor* or *divisor* of the other.

We use *divides* here to mean 'divides exactly and leaves no remainder'.

The main aim of this section is to prove the following result.

Theorem 1.1 Quotient–remainder theorem (special case)

Let a and b be integers with $b > 0$.
Then there exist unique integers q and r such that

$$a = bq + r,$$

where r satisfies

$$0 \leq r < b.$$

We say that q is the **quotient** and r is the **remainder** on dividing a by b.

We have called this a special case because we have imposed the restriction that b is positive. We shall prove a more general form of the theorem later.

The proof strategy is based on rearranging $a = bq + r$ as

$$a - bq = r.$$

This rearrangement leads us to consider all integers of the form

$$a - b \times \text{(some integer)},$$

and to show that one of these has the properties required of r.

Underlying the proof, and many of the other results that we shall prove, is a basic assumption about the positive integers $\mathbb{N}$. Since this property of $\mathbb{N}$ will be used frequently, we state it here and give it its usual name.

Theorem 1.2 Well-ordering property of $\mathbb{N}$

Let S be a non-empty subset of $\mathbb{N}$. Then S has a least element.

The set $\mathbb{N}$ is an *ordered set*. This result says that the ordering has additional properties, hence the name. $\mathbb{Z}$ is ordered, but is *not* well-ordered: consider, for example, the subset $\mathbb{Z}$ itself, which does not have a least element.

We shall not give a proof of this theorem. To do so would require a formal definition of $\mathbb{N}$, which is not appropriate for this course. This theorem is sometimes used in proofs as an alternative to the Principle of Mathematical Induction.

We have followed the usual convention that 0 is not in $\mathbb{N}$. However, Theorem 1.2 is still true if $\mathbb{N}$ is replaced by the set of non-negative integers. This is because, if S is a non-empty subset of the non-negative integers and contains 0, then 0 is the least element of S. On the other hand, if S does not contain 0, then S is a non-empty subset of $\mathbb{N}$ and Theorem 1.2 applies directly. We shall sometimes exploit this extended version which says that the non-negative integers are well-ordered.

We can derive another useful result from the well-ordering property of $\mathbb{N}$. But, before we can state it, we need a definition.

Definition 1.1

A subset S of $\mathbb{Z}$ is **bounded below** if there exists an integer m such that $m \leq s$ for all $s \in S$, and is **bounded above** if there exists an integer M such that $s \leq M$ for all $s \in S$. The subset S is **bounded** if it is both bounded below and bounded above.

We refer to m as a **lower bound** for S and to M as an **upper bound**.

Theorem 1.3

Let S be a non-empty subset of $\mathbb{Z}$.
(a) If S is bounded below, then S has a least element.
(b) If S is bounded above, then S has a greatest element.

Proof

(a) S is bounded below, by the integer m say, so that
$$m \leq s, \quad \text{for all } s \in S.$$
Hence
$$0 \leq s - m, \quad \text{for all } s \in S,$$
and thus the set
$$T = \{s - m : s \in S\}$$
consists of non-negative integers.
Therefore, by the extended version of Theorem 1.2, T has a least element, l_T say.

We know that $l_T \in T$. Hence
$$l_T = l_S - m, \quad \text{for some } l_S \in S.$$
Furthermore,
$$l_T \leq m - s, \quad \text{for all } s \in S.$$
Therefore
$$l_S - m \leq s - m, \quad \text{for all } s \in S,$$
giving
$$l_S \leq s, \quad \text{for all } s \in S.$$
Hence $l_S \in S$ and $l_S \leq S$ for all $s \in S$.
Thus l_S is the least element of S. □

Exercise 1.1

Prove part (b) of Theorem 1.3.

Proof of Theorem 1.3 continued

Hence the theorem is proved. ∎

Let us now move on to the proof of our main result of this section, Theorem 1.1. Our proof is in two parts, corresponding to the two assertions of the theorem: existence and uniqueness.

Proof of Theorem 1.1

Existence

The proof of existence is in two parts, corresponding to $a \geq 0$ and $a < 0$.

We begin with the case $a \geq 0$.

Following our earlier comment about proof strategy, we consider the set
$$S = \{a - bz : z \in \mathbb{Z}, a - bz \geq 0\}.$$
Firstly, S is non-empty, because the integer
$$a - b \times 0 = a$$
is of the right form, is non-negative and so is in S.
Secondly, by definition, S is a subset of the non-negative integers.
Therefore, by the extended well-ordering property, S has a least element.

The way we define S ensures that, if it contains any elements at all, then they will be non-negative. We aim to apply the extended well-ordering property to S.

Suppose that the least element of S is r and the corresponding value of z is q. In other words,
$$a - bq = r.$$
So we have shown the existence of q and r such that
$$a = bq + r.$$
To complete the existence proof for the case $a \geq 0$, we must show that
$$0 \leq r < b.$$
From the definition of S, we know that $0 \leq r$, so we just have to show that $r < b$. We do this by contradiction.
Suppose that $r \not< b$, that is
$$b \leq r.$$
We observe that, because b is positive,
$$r < r + b.$$
Thus, we have
$$b \leq r < r + b.$$
Subtracting b throughout, gives
$$0 \leq r - b < r.$$
So $r - b$ is a non-negative integer. Furthermore,
$$\begin{aligned}r - b &= (a - bq) - b \\ &= a - b(q+1).\end{aligned}$$
This implies that $r - b$ is a non-negative integer of the form $a - bz$ and so is an element of S.
On the other hand,
$$0 \leq r - b < r$$
contradicts the choice of r as the least element of S.
Hence, as required, $r < b$.
We have now established the *existence* of q and r satisfying
$$a = bq + r, \quad 0 \leq r < b,$$
in the case $a \geq 0$.

In the following exercise we ask you to deal with the case $a < 0$. □

Exercise 1.2

Given integers $a < 0$ and $b > 0$, show that there exist integers $\tilde{q}$ and $\tilde{r}$ such that
$$a = b\tilde{q} + \tilde{r}, \quad 0 \leq \tilde{r} < b.$$

Hint Since $|a| = -a$ is positive, you can apply the case proved above to $|a|$ and b. Note also that, if
$$0 < r < b,$$
then
$$0 < b - r < b.$$

Proof of Theorem 1.1 continued

We now complete the proof of the theorem by dealing with uniqueness.

Uniqueness

Suppose that q_1 and r_1 and also q_2 and r_2 satisfy the requirements of Theorem 1.1. That is:
$$a = bq_1 + r_1, \quad 0 \leq r_1 < b;$$
$$a = bq_2 + r_2, \quad 0 \leq r_2 < b.$$
Subtracting these gives
$$0 = b(q_1 - q_2) + (r_1 - r_2). \tag{1.1}$$

As usual in 'uniqueness' proofs, we assume that there are two expressions for a of the desired form and show that they must be the same.

We proceed by contradiction.

If the qs are not equal, we may assume that $q_1 > q_2$. Hence
$$q_1 - q_2 > 0,$$
and, therefore,
$$q_1 - q_2 \geq 1.$$
From Equation 1.1, since $b > 0$, we have
$$\begin{aligned}r_2 - r_1 &= b(q_1 - q_2) \\ &\geq b \times 1 = b.\end{aligned}$$
It follows that
$$r_2 \geq r_1 + b \geq b.$$
This contradicts the assumption that $r_2 < b$. Therefore our assumption that the qs are not equal cannot be true, and so $q_1 = q_2$. It follows immediately from Equation 1.1 that $r_1 = r_2$, and we are finished. ∎

We now ask you to extend Theorem 1.1 to the case where $b < 0$. This will give us the final form of the Quotient–remainder Theorem.

Exercise 1.3

Given any integer a and an integer $b < 0$, show that there exist unique integers q and r such that
$$a = bq + r, \quad 0 \leq r < |b|.$$

Hint Apply the results proved so far to a and $|b|$.

Combining Theorem 1.1 with the result of Exercise 1.3 gives the following.

Theorem 1.4 Quotient–remainder theorem (general form)

Let a and b be integers with $b \neq 0$.
Then there exist unique integers q and r such that
$$a = bq + r,$$
where r satisfies
$$0 \leq r < |b|.$$

For example:
$$\begin{aligned}19 &= 3 \times 6 + 1; \\ -13 &= (-3) \times 5 + 2; \\ 11 &= (-4) \times (-2) + 3; \\ -19 &= 4 \times (-5) + 1.\end{aligned}$$

The special case in which the remainder is zero, and we have exact division in $\mathbb{Z}$, is of some importance. In such a case
$$a = bq + r, \quad 0 \leq r < |b|,$$
reduces to
$$a = bq.$$
We formalize this situation in the following definition.

Definition 1.2 Divides

Let a and b be integers. We say that b **divides** a, written
$$b \mid a,$$
if there is an integer q such that
$$a = bq.$$
We also say that b is a **factor** or **divisor** of a and that a is a **multiple** of b.

Unlike Theorem 1.4, the definition of divisibility allows the case $b = 0$.

Example 1.1

Every integer divides zero. If b is any integer, then

$$0 = b \times 0.$$ ♦

Example 1.2

Every integer divides itself. If a is any integer, then

$$a = a \times 1.$$ ♦

The following exercises establish some other useful consequences of the definition of divisibility. All you will need to construct the proofs is the definition of divisibility.

Exercise 1.4

Prove that the only integer divisible by zero is zero.

Exercise 1.5

For integers x, y and z, prove that if

$$x \mid y \quad \text{and} \quad y \mid z$$

then

$$x \mid z.$$

Exercise 1.6

Suppose that a, b and c are integers such that

$$c \mid a \quad \text{and} \quad c \mid b.$$

Prove that

$$c \mid (\alpha a + \beta b),$$

for all integers α and β.

We can restate the result from Exercise 1.6 in an alternative form by first making the following definition.

Definition 1.3 Integer combination

Let a and b be integers. Then any integer of the form

$$\alpha a + \beta b, \quad \alpha, \beta \in \mathbb{Z},$$

is called an **integer combination** of a and b.

With this definition, the result of Exercise 1.6 can be stated as follows.

Lemma 1.1

Let a, b and c be integers such that c divides both a and b.
Then c divides any integer combination of a and b.

As we said in the Introduction, the reason we prove results such as the ones in this section is to obtain results about groups. As an example of an application, we apply Theorem 1.4 to prove the following result about orders of group elements.

> **Lemma 1.2**
>
> Let g be an element of the group G. If g has order n, then
>
> $$g^m = e \iff n \mid m.$$

The symbol $\iff$ means 'if and only if'.

Proof

If

Since g has order n, we have $g^n = e$.
If $n \mid m$ then there is an integer q such that

$$m = nq.$$

Hence

$$\begin{aligned} g^m &= g^{nq} \\ &= (g^n)^q \\ &= e^q \\ &= e. \end{aligned}$$

Only if

Since the order of an element is non-zero, $n \neq 0$, and we can apply Theorem 1.4 to write

$$m = nq + r, \quad 0 \leq r < n.$$

We are given two pieces of information. Firstly, $g^m = e$. Secondly, since g has order n, n is the *smallest* positive integer satisfying $g^n = e$.

Now,

$$\begin{aligned} g^r &= g^{m-nq} \\ &= g^m g^{-nq} \\ &= e g^{-nq} \quad \text{(since } g^m = e\text{)} \\ &= (g^n)^{-q} \\ &= e^{-q} \quad \text{(since } g^n = e\text{)} \\ &= e. \end{aligned}$$

This shows that $g^r = e$.

Since n is the smallest *positive* integer for which this is true, and since $r < n$, r cannot be positive. Hence we have

$$r \leq 0.$$

Combining this with the fact that $0 \leq r$ gives $r = 0$. Hence

$$n \mid m. \qquad \blacksquare$$

Note that the proof above used a common strategy for showing that one integer divides another. We first used the Quotient–remainder Theorem and then showed that the remainder had to be zero. In order to show that the remainder was zero, we used the fact that the divisor was the smallest *positive* integer with a particular property, showed that the remainder possessed the same property and deduced that the remainder could not be positive.

2 FACTORS AND MULTIPLES

In this section we define highest common factors and lowest common multiples and prove a series of results which will come in useful when discussing finite groups.

> **Definition 2.1 Highest common factor**
>
> Let a and b be integers, not both zero. Then the **highest common factor** (HCF) of a and b is the largest integer dividing both a and b.
>
> We denote the highest common factor of two integers a and b (not both zero) by
>
> $\mathrm{hcf}\{a, b\}$.

Highest common factors are sometimes referred to as greatest common divisors.

A **common factor** of two integers a and b is an integer c that is a factor of both a and b.

Whenever discussing HCFs, we shall assume that at least one of the integers concerned is non-zero. As this is required by the definition, we may not always state this assumption explicitly.

The set notation for HCFs is justified because the definition of HCF is symmetrical, that is

$\mathrm{hcf}\{a, b\} = \mathrm{hcf}\{b, a\}$.

Two simple examples of highest common factors are that 3 is the HCF of 6 and 15 and that 1 is the HCF of -5 and 4.

The definition of highest common factors raises the question of whether they always exist.

To answer this, we first note that *common* factors of any two integers a and b always exist, because the two equations

$a = a \times 1$ and $b = b \times 1$,

show that 1 is always a common factor of any two integers.

Secondly, there is an upper bound to the size of factors (and hence common factors) of any two integers a and b, not both zero. To see why, observe that since a and b are not both zero, and since the definition of HCF is symmetric in a and b, we may assume that $a \neq 0$. Now suppose that c is a factor of a so that

$a = cq$

for some $q \in \mathbb{Z}$. Then

$|a| = |c| |q|$.

As $a \neq 0$, we have $q \neq 0$, so $|q| \geq 1$ and hence

$|a| = |c| |q| \geq |c| \times 1 = |c|$.

Hence, for any factor c of a, we have $|c| \leq |a|$. Since $c \leq |c|$, it follows that

$c \leq |a|$.

Similarly, if $b \neq 0$ then, for any factor c of b, we have

$c \leq |b|$.

Therefore, if a and b are both non-zero then any common factor cannot exceed the smaller of $|a|$ and $|b|$. To allow for one of a and b being zero, however, we can only make the weaker statement that any common factor c of a and b cannot exceed the larger of $|a|$ and $|b|$.

We now know that common factors exist and have an upper bound. We can deduce, using Theorem 1.3(b), that highest common factors always exist.

You may find other notation, such as (a, b), used instead of $\mathrm{hcf}\{a, b\}$ in some textbooks. We do not use the notation (a, b) in order to avoid confusion with ordered pairs.

In the process of proving the existence of HCFs above, we have proved a general result about division which will come in useful in later proofs. The result is as follows.

> **Lemma 2.1**
>
> Let a and b be integers with $a \neq 0$. If b divides a, then
>
> $b \leq |b| \leq |a|$.

There are several facts about HCFs that arise directly from the definition.

Because, as we saw above, there is always at least one *positive* common factor, namely 1, HCFs are always positive.

Because every integer divides zero, the definition shows that

$\text{hcf}\{a, 0\} = a, \quad \text{for all } a > 0.$

Exercise 2.1

Let a and b be non-zero integers. Show that

$\text{hcf}\{a, b\} = \text{hcf}\{a, |b|\} = \text{hcf}\{|a|, b\} = \text{hcf}\{|a|, |b|\}.$

Hint Consider the factors of a and $|a|$ and of b and $|b|$.

As a consequence of Exercise 2.1, and the remark preceding it,

$\text{hcf}\{a, 0\} = |a|, \quad \text{for all } a \neq 0.$

Because of this result, we shall often restrict ourselves to discussing HCFs of non-zero integers.

The HCF of two integers has one particular property that is not immediately obvious from the definition but which is extremely useful.

> **Theorem 2.1**
>
> Let a and b be integers, not both zero, and let $d = \text{hcf}\{a, b\}$. Then d is the smallest, positive, integer combination of a and b.

Proof

We consider the set S of *all*, positive, integer combinations of a and b. That is,

$S = \{xa + yb : x, y \in \mathbb{Z}, \ xa + yb > 0\}.$

We first observe that the set S is not empty.

As a and b are not both zero, and the definition of HCF is symmetric, we may assume that $a \neq 0$. Hence one of the integer combinations, namely

$\pm 1a + 0b,$

is $|a|$, which is positive and hence in S.

Since S is non-empty, by the well-ordering property of $\mathbb{N}$ it has a least element. Let this least element be d, given by the integer combination

$d = \alpha a + \beta b,$

for suitable integers α and β.

Next we show that d is a common factor of a and b. We start with a.

Since $d \in S$, $d \neq 0$, and so by Theorem 1.4 we can write

$$a = dq + r, \quad 0 \leq r < |d| = d.$$

We need to show that $r = 0$.

Substituting for d in this expression for a we get

$$a = dq + r$$
$$= (\alpha a + \beta b)q + r.$$

Hence

$$r = a - (\alpha a + \beta b)q$$
$$= (1 - \alpha q)a + (-\beta q)b.$$

This shows that r is an integer combination of a and b.

Now, by definition, r is non-negative and is strictly less than d, i.e.

$$0 \leq r < d.$$

If r were positive, it would be in S, contradicting the minimality of d. Hence $r = 0$ and so d divides a.

A parallel argument shows that d divides b.

Hence d is a common factor of a and b.

Finally, we must show that d is the *largest* common factor of a and b; in other words that any common factor is less than or equal to d.

Suppose that c is a common factor of a and b. Since d is an integer combination of a and b, it follows from Lemma 1.1 that

$$c \mid d$$

and, therefore, by Lemma 2.1,

$$c \leq |d| = d.$$

This completes the proof. ∎

> We use the Quotient–remainder Theorem and then show that the remainders are zero.

There are two useful results about HCFs which are contained in the proof of Theorem 2.1.

(a) The HCF of two integers is not only greater than any other common factor but also *divisible by* any such common factor.

(b) A positive common factor which is divisible by *all* common factors must be the highest common factor.

In fact, we could have defined an HCF to be a positive common factor that is divisible by all common factors.

A further useful result is that $\mathrm{hcf}\{a, b\}$ divides any integer combination of a and b.

The special case where the highest common factor of a and b is 1 is of particular interest. Hence we make the following definition.

Definition 2.2 Coprime

If a and b are integers such that

$$\mathrm{hcf}\{a, b\} = 1,$$

then we say that a and b are **coprime**.

> Because coprimeness is defined in terms of HCF, we are assuming that a and b are not both zero.

Exercise 2.2

Show that integers a and b are coprime if and only if there exist integers α and β such that

$$\alpha a + \beta b = 1.$$

Exercise 2.3

Let a and b be integers and $d = \text{hcf}\{a, b\}$. Write $a = dx$ and $b = dy$.
Show that x and y are coprime.

Exercise 2.4

Let a, b and c be integers such that a divides bc and $\text{hcf}\{a, b\} = 1$.
Show that a divides c.

Exercise 2.5

Let a, b and c be integers such that a and b divide c and $\text{hcf}\{a, b\} = 1$.
Show that ab divides c.

Our work so far in this section has concentrated on the existence of HCFs, the existence of integer combination forms for HCFs and some related results. In the next section of this unit we shall give an algorithm for calculating HCFs. However, a number of results about groups depend only on the existence of HCFs and on the existence of lowest common multiples, which we now define.

Definition 2.3 Lowest common multiple

Let a and b be non-zero integers. Then the **lowest common multiple** (LCM) of a and b is the smallest positive integer divisible by both a and b. In other words, the lowest common multiple is the smallest positive common multiple of a and b.

We denote the lowest common multiple of two non-zero integers a and b by

$$\text{lcm}\{a, b\}.$$

Unlike the case of HCFs, *both* must be non-zero.

A **common multiple** of two integers a and b is an integer c that is a multiple of both a and b.

The proofs of statements about LCMs are often similar to the proofs of the corresponding statements for HCFs. We ask you to prove two of these statements in the following exercises.

Exercise 2.6

Show that the LCM of two non-zero integers a and b exists.

Exercise 2.7

Let a and b be non-zero integers and let $m = \text{lcm}\{a, b\}$.
If n is any common multiple of a and b, show that m divides n.

Hint Use the Quotient–remainder Theorem and then show that the remainder is zero.

Although we do not prove it here, the following result is also true:

$$\text{lcm}\{a, b\} = \text{lcm}\{a, |b|\} = \text{lcm}\{|a|, b\} = \text{lcm}\{|a|, |b|\}.$$

The connection between HCFs and LCMs is given in the following theorem.

> **Theorem 2.2**
>
> Let a and b be non-zero integers. Then
> $$\mathrm{hcf}\{a,b\} \times \mathrm{lcm}\{a,b\} = |ab|.$$

Proof
Because of the observation before the theorem, it suffices to prove that for *positive* a and b

$$\mathrm{hcf}\{a,b\} \times \mathrm{lcm}\{a,b\} = ab.$$

Let $d = \mathrm{hcf}\{a,b\}$, $m = \mathrm{lcm}\{a,b\}$, with $a = dx$ and $b = dy$.

> Since d is positive by definition and since a and b are positive by assumption, x and y must be positive too.

We wish to prove that

$$dm = (dx)(dy);$$

in other words that

$$m = dxy.$$

Since

$$dxy = (dx)y = ay$$

and

$$dxy = (dy)x = bx,$$

we have that dxy is a common multiple of both a and b.
Hence, by Exercise 2.7, m divides dxy.

On the other hand, by Theorem 2.1, we can write

$$d = \alpha a + \beta b = \alpha dx + \beta dy.$$

So, dividing through by d, we have

$$1 = \alpha x + \beta y.$$

Multiplying this throughout by m we obtain

$$m = \alpha x m + \beta y m. \tag{2.1}$$

Now, $dx = a$ divides m, and so dxy divides ym. Also $dy = b$ divides m, and so dxy divides xm.
By Equation 2.1, m is an integer combination of xm and ym, so it follows, by Exercise 1.6, that

$$dxy \mid m.$$

Since all of a, b, d, m, x and y are positive,

$$m \mid dxy \quad \text{and} \quad dxy \mid m$$

give, by Lemma 2.1,

$$m \leq dxy \quad \text{and} \quad dxy \leq m.$$

Hence $m = dxy$, and so $dm = dxdy = ab$ as required. ∎

In the following exercises we ask you to apply the ideas of this section to direct products of groups.

Exercise 2.8

This exercise concerns the direct product $\mathbb{Z}_4 \times \mathbb{Z}_6$.

(a) Write down the orders of $2 \in \mathbb{Z}_4$, $2 \in \mathbb{Z}_6$ and $(2,2) \in \mathbb{Z}_4 \times \mathbb{Z}_6$.

(b) Repeat part (a) to find the orders of $a \in \mathbb{Z}_4$, $b \in \mathbb{Z}_6$ and $(a,b) \in \mathbb{Z}_4 \times \mathbb{Z}_6$ for all 24 possible choices of a and b.

From the results of the previous exercise, you might guess that the order of an element in the direct product is the LCM of the orders of its components. We now ask you to show that this is true in general.

Exercise 2.9

Let G and H be finite groups with elements $g \in G$ and $h \in H$. If g has order m and h has order n, show that the order of the element (g, h) in the direct product $G \times H$ is $\text{lcm}\{m, n\}$.

3 THE EUCLIDEAN ALGORITHM (AUDIO-TAPE SECTION)

In the previous section we proved the existence and uniqueness of highest common factors. We now discuss a practical method of calculating them. Because we know, from Theorem 2.2, that

$$\text{hcf}\{a, b\} \times \text{lcm}\{a, b\} = |ab|,$$

this method will also produce LCMs.

You may already have met a method of finding the HCF of two integers a and b that depends on factorizing them into primes. However, this factorization method is not very satisfactory since it requires a relatively large number of arithmetic operations. The method we describe here, known as the *Euclidean Algorithm*, usually requires rather fewer operations. Furthermore, because it is an *algorithm*, i.e. each step of the method is predetermined, the whole process is capable of being automated.

We shall discuss primes in Section 4 of this unit.

You should now listen to the audio programme for this unit, referring to the tape frames below when asked to do so during the programme.

1 Overview

Aim
Find a method for calculating highest common factors 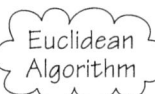 (Euclidean Algorithm)

Definition
The highest common factor of a and b is the largest integer dividing them both (hcf$\{a,b\}$) (a, b can't both be 0)

Quotient–remainder Theorem
If a and b are integers and b is positive, then there exist unique integers q and r such that
$a = bq + r$, $0 \leq r < b$ (use this result to justify method)

2 Simplifications

(a) hcf$\{a,0\}$ = $|a|$ (easy if either number is zero)

(b) hcf$\{a,b\}$ = hcf$\{b,a\}$ (order doesn't matter)

(c) hcf$\{a,b\}$ = hcf$\{a,|b|\}$ = hcf$\{|a|, b\}$ = hcf$\{|a|,|b|\}$ (remove negatives)

(d) hcf$\{a,a\}$ = a, $a > 0$ (assume not equal)

Simplified task: find hcf$\{a,b\}$ where $a > b > 0$

3 Examples

(a) hcf$\{12,0\}$ = 12 (hcf$\{a,0\}$ = $|a|$)

(b) hcf$\{0,-16\}$ = hcf$\{0,16\}$ = 16 (remove negatives)

(c) hcf$\{-68,-13\}$ = hcf$\{68,13\}$
 then use algorithm (hcf$\{a,b\}$ = hcf$\{|a|,|b|\}$)

(d) hcf$\{-15, 79\}$ = hcf$\{15,79\}$
 = hcf$\{79,15\}$ (change order to get $a > b$)
 then use algorithm

4 Strategy

Overall Reduce hcf$\{a,b\}$ to hcf$\{x,0\}$ = x ($a > b > 0$) ($x > 0$)

Reduction step Replace a, b with a', b' where hcf$\{a', b'\}$ = hcf$\{a,b\}$ and $b' < b$ ($a' > b' \geq 0$)  (bs decrease)

Repeat until a zero appears

5

The process of the reduction step

Replace a, b ($a > b$) with $a' = b$
$b' = a \bmod b$ (remainder on division by b)

Example

$$\text{hcf}\{128,60\} = \text{hcf}\{60,8\} \quad (128 = 60 \times 2 + \boxed{8})$$
$$= \text{hcf}\{8,4\} \quad (60 = 8 \times 7 + \boxed{4})$$
$$= \text{hcf}\{4,0\} \quad (8 = 4 \times 2 + \boxed{0})$$
$$= 4$$

6

Exercise 3.1

Use the method of the previous frame to find:

(a) hcf{144,128}
(b) hcf{68,13}
(c) hcf{157,28}

6A

Solution 3.1

(a) $\text{hcf}\{144,128\}$
 $= \text{hcf}\{128,16\}$ ($144 = 128 \times 1 + 16$)
 $= \text{hcf}\{16,0\}$ ($128 = 16 \times 8 + 0$)
 $= 16$

(b) $\text{hcf}\{68,13\}$
 $= \text{hcf}\{13,3\}$ ($68 = 13 \times 5 + 3$)
 $= \text{hcf}\{3,1\}$ ($13 = 3 \times 4 + 1$)
 $= \text{hcf}\{1,0\}$ ($3 = 1 \times 3 + 0$)
 $= 1$

(c) $\text{hcf}\{157,28\}$
 $= \text{hcf}\{28,17\}$ ($157 = 28 \times 5 + 17$)
 $= \text{hcf}\{17,11\}$ ($28 = 17 \times 1 + 11$)
 $= \text{hcf}\{11,6\}$ ($17 = 11 \times 1 + 6$)
 $= \text{hcf}\{6,5\}$ ($11 = 6 \times 1 + 5$)
 $= \text{hcf}\{5,1\}$ ($6 = 5 \times 1 + 1$)
 $= \text{hcf}\{1,0\}$ ($5 = 1 \times 5 + 0$)
 $= 1$

7 The reduction step formalized

Given $\quad a > b > 0 \quad$ (Does $\text{hcf}\{a,b\} = \text{hcf}\{a',b'\}$?)

Reduce $\begin{cases} a' = b \\ b' = a \bmod b \\ \text{where } a = bq + r,\ 0 \leq r < b \\ \text{so } b' = r \end{cases}$ (Quotient–remainder Theorem)

New pair $\quad \begin{array}{cc} a' & b' \\ \downarrow & \downarrow \\ b & > r \geq 0 \end{array}$ (automatically get $b' = r < b$ and bs reduce to zero)

Want $\quad \text{hcf}\{a,b\} = \text{hcf}\{b,r\} = \text{hcf}\{a',b'\}$

8 HCFs are equal

Strategy Show all common divisors of a and b are common divisors of b and r and conversely

(common divisors = common factors)

Proof (a) Have $a = bq + r$
If x is a common divisor of b and r (integer combination)
then x divides $bq + r = a$
So x is common divisor of a and b

(converse) (b) Have $r = a - bq$
If y is a common divisor of a and b (integer combination)
then y divides $a - bq = r$
So y is common divisor of b and r

so $\text{hcf}\{a,b\} = \text{hcf}\{b,r\} = \text{hcf}\{a',b'\}$

9 The complete algorithm

Each reduction step reduces the remainder, so the process must stop ($0 \leq r < b$)

$a = bq_1 + r_1 \qquad \text{hcf}\{a,b\} = \text{hcf}\{b,r_1\}$
$b = r_1 q_2 + r_2 \qquad \text{hcf}\{b,r_1\} = \text{hcf}\{r_1,r_2\}$
$r_1 = r_2 q_3 + r_3 \qquad \text{hcf}\{r_1,r_2\} = \text{hcf}\{r_2,r_3\}$

$\vdots$

$r_{n-3} = r_{n-2} q_{n-1} + r_{n-1} \quad \text{hcf}\{r_{n-3},r_{n-2}\} = \text{hcf}\{r_{n-2},r_{n-1}\}$
$r_{n-2} = r_{n-1} q_n \qquad \text{hcf}\{r_{n-2},r_{n-1}\} = \text{hcf}\{r_{n-1},0\} = r_{n-1}$

(at most b steps)

10

Integer combinations

hcf{128,60} = 4 — *HCF is an integer combination of 128 and 60*

$128 = 60 \times 2 + 8$ $8 = 128 - 60 \times 2$
$60 = 8 \times 7 + 4$ $4 = 60 - 8 \times 7$
$8 = 4 \times 2 + 0$

Substituting (last non-zero remainder ← ; previous remainder →)

$$4 = 60 - 8 \times 7$$
$$= 60 - (128 - 60 \times 2) \times 7$$
$$= 60 \times 15 - 128 \times 7$$

$$4 = (-7) \times 128 + 15 \times 60$$

11

Exercise 3.2

Express each of these HCFs as linear combinations:

(a) hcf{144,128}

(b) hcf{68,13}

you've already found the HCFs in Frame 6

11A

Solution 3.2

(a) $144 = 128 \times 1 + 16$ $16 = 144 - 128 \times 1$
     $128 = 16 \times 8 + 0$

hcf{144,128} = $16 = 1 \times 144 + (-1) \times 128$

(b) $68 = 13 \times 5 + 3$ $3 = 68 - 13 \times 5$
     $13 = 3 \times 4 + 1$ $1 = 13 - 3 \times 4$
     $3 = 1 \times 3 + 0$

$$\text{hcf}\{68,13\} = 1 = 13 - 3 \times 4$$
$$= 13 - (68 - 13 \times 5) \times 4$$
$$= 13 - 68 \times 4 + 13 \times 20$$
$$= (-4) \times 68 + 21 \times 13$$

12

Summing up

To calculate hcf{a,b}

(a) Simplify to non-negative numbers, and reorder so $a > b$

(b) Reduce the larger modulo the smaller. Continue until a zero appears
    ($a \to a' = b$; $b \to b' = a \bmod b$)

(c) hcf{a,b} is the last non-zero remainder

(d) To express an HCF as an *integer combination*, write out the divisions in full and back substitute for the remainders

13 Thoughts on the general strategy

Stage 1 Identify easy cases

Stage 2 Find a method of reducing all other cases to one of the easy ones

> This is a very powerful general strategy

For convenience, we set out the Euclidean Algorithm here.

Euclidean algorithm

The HCF of non-zero integers a and b, where $|a| > |b|$, is found by performing the following calculations:

$$|a| = |b|q_1 + r_1, \quad 0 \leq r_1 < |b|;$$
$$|b| = r_1 q_2 + r_2, \quad 0 \leq r_2 < r_1;$$
$$r_1 = r_2 q_3 + r_3, \quad 0 \leq r_3 < r_2;$$
$$\vdots$$
$$r_{n-3} = r_{n-2} q_{n-1} + r_{n-1}, \quad 0 \leq r_{n-1} < r_{n-2};$$
$$r_{n-2} = r_{n-1} q_n.$$

Then $\text{hcf}\{a, b\}$ is r_{n-1}, the last non-zero remainder.

Exercise 3.3

(a) Use the Euclidean Algorithm to show that 111 and 421 are coprime.

(b) Use the result of part (a) to express $1 = \text{hcf}\{111, 421\}$ as an integer combination of 111 and 421.

Exercise 3.4

Using the value of $\text{hcf}\{144, 128\}$ obtained in the solution to Exercise 3.1 (Frame 6A), find $\text{lcm}\{144, 128\}$.

4 PRIMES

The aims of this section, and the next, are to formalize ideas of exactly what prime numbers are and to show that every integer greater than 1 has a unique factorization into a product of prime numbers.

Definition 4.1 Prime number

A positive integer p is a **prime** number if it has exactly two positive factors.

We shall often shorten 'p is a prime number' to 'p is prime' or to 'p is a prime'.

The following are immediate consequences of the definition.

(a) Since every integer is divisible both by itself and by 1, the two positive factors of a prime p are p and 1.

(b) Every prime p has exactly four factors: $1, -1, p, -p$.

(c) Since 1 has exactly one positive factor, 1 is *not* prime.

The following theorem illustrates an application of the definition of prime number to group theory.

> **Theorem 4.1**
>
> Let p be a prime number. Then there is only one group of order p, namely the cyclic group of order p.

Exercise 4.1

Use the definition of prime number, and Lagrange's Theorem, to prove Theorem 4.1.

At first sight, testing a given positive integer n for being prime involves checking whether it is divisible by any of the integers

$$2, \ldots, n-1.$$

However, if $n = n_1 n_2$ then one of the factors must be less than or equal to $\sqrt{n}$, since if both n_1 and n_2 are greater than $\sqrt{n}$ their product is greater than n. Thus we need only check for factors up to $\sqrt{n}$. Devising efficient tests for prime numbers is a hard problem and we shall not be concerned with such testing.

You will find it useful to be able to recognize the first few primes, which are

$$2, 3, 5, 7, 11, 13, 17, 19, 23, 29, 31, \ldots$$

If p is prime we cannot write p as a product of positive integers both of which are less than p. On the other hand, if an integer n greater than 1 is not prime, then this is exactly what we *can* do. This is a consequence of the definition, as the formal proof of the following lemma shows.

> **Lemma 4.1**
>
> Let n be an integer greater than 1 that is not prime.
> Then n can be expressed as the product of two positive integers both of which are greater than 1 and less than n.

Proof

Since n is not prime, it has a positive factor n_1, say, which is neither 1 nor n. Hence

$$n = n_1 n_2,$$

say. Since n_1 is a factor of n we have

$$n_1 = |n_1| \leq |n| = n.$$

A reminder: $b \mid a \Rightarrow |b| \leq |a|$.

Since $n_1 \neq n$, we have

$$n_1 < n.$$

Also, since n_1 is positive and not equal to 1,

$$1 < n_1.$$

Now look at n_2. Since $n_1 \neq n$, we know that $n_2 \neq 1$. Similarly, $n_1 \neq 1$ and so $n_2 \neq n$. Thus n_2 is a positive factor of n which is neither n nor 1. We argue as above to obtain

$$1 < n_2 < n. \qquad \blacksquare$$

In the above situation, we can say more: n must actually have a *prime* factor, i.e. a factor that is a prime number. More formally, we have the following.

> ### Theorem 4.2
> Let n be an integer greater than 1. Then n has a prime factor.

This assertion is the first step towards expressing n as a product of primes.

Our proof of this theorem is an example of an important proof strategy called the *minimal counterexample* strategy. It is a variant of proof by contradiction. The reason for the name should become apparent during the proof.

Proof

Assume that the theorem is not true. Then there is a counterexample; that is, there is an integer greater than 1 which has no prime factor. Thus we are supposing that the set of counterexamples is non-empty. Since the set of such counterexamples is a subset of $\mathbb{N}$, it has a least element, n say.

This is the reason for calling it the 'minimal counterexample' strategy.

Since n has no prime factor, it cannot be prime (otherwise n would be a prime factor of itself) and therefore, by Lemma 4.1, can be written

$$n = n_1 n_2,$$

where n_1 is *less* than n and greater than 1. By the choice of n as minimal counterexample, n_1 *does* have a prime factor, p say.

However, we have that p divides n_1 and n_1 divides n, which imply that p divides n, a contradiction.

See Exercise 1.5.

This contradiction shows that the theorem must be true. ∎

It is useful to have a name for non-prime integers which are greater than 1.

> ### Definition 4.2 Composite
> A non-zero, non-prime integer n is a **composite** integer if $n \neq \pm 1$.

There are other methods of proving Theorem 4.2. The next exercise asks you to provide an alternative proof.

Exercise 4.2

Suppose that n is a positive composite integer. By considering the set of all positive factors of n other than itself and 1, prove that n has a prime factor.

We now have all the tools that we need to show that every integer greater than 1 can be expressed as a product of primes.

We regard a prime as a 'product of one prime'.

> ### Theorem 4.3
> Every integer greater than 1 can be expressed as a product of primes.

Proof

We use the minimal counterexample strategy again. Suppose that the theorem is not true and that n is the smallest integer greater than 1 that *cannot* be expressed as a product of primes.

Now, n cannot be prime, otherwise, by our convention it would be a product of primes. It follows that n is a positive composite integer and so, by Theorem 4.2, it has a prime factor, p_1 say. That is, we can write

$$n = p_1 n_1,$$

where p_1 is prime.

Now $p_1 \neq 1$ and $p_1 \neq n$, hence (as in the proof of Lemma 4.1)

$$1 < n_1 < n.$$

By the minimality of n, we know that n_1 *can* be expressed as a product

$$n_1 = p_2 p_3 \ldots p_k$$

of primes $p_2, \ldots, p_k$.

Substituting for n_1 gives

$$n = p_1 p_2 \ldots p_k,$$

which contradicts the assumption about n.

This contradiction completes the proof of the theorem. ∎

As you may have guessed, there is an alternative, direct proof, of Theorem 4.3 using the Principle of Mathematical Induction.

The various results that we have obtained suggest a method of expressing a given positive composite integer, n, as a product of primes. We can deduce from the proof of Theorem 4.2 that the smallest factor of n greater than 1 is prime. So we check the primes, starting with the smallest. Once we find a factor, we divide out and repeat with the quotient.

Exercise 4.3

Express 1728 as a product of primes.

It is difficult to see how any method of expressing 1728 as a product of primes could possibly lead to a different product of prime factors apart, perhaps, from the order in which the factors are written. If we agree to write the primes in ascending order, then it would appear that the factorization of any integer greater than 1 into a product of primes must be unique.

In fact, every integer greater than 1 *can* be written uniquely as a product of primes in ascending order. We have established the existence of such a factorization in Theorem 4.3; the uniqueness will be proved in the next section.

In the meantime, we ask you to prove some preliminary results.

Exercise 4.4

Suppose that p is prime and a is a non-zero integer such that p does not divide a. Show that p and a are coprime.

Exercise 4.5

Prove that any two distinct primes p and q are coprime.

Exercise 4.6

Suppose that a, b and c are integers, that a and b are coprime and that b and c are coprime. Is it true that a and c must be coprime? Give either a proof or a counterexample as appropriate.

5 UNIQUE PRIME FACTORIZATION

The main aim of this section is to prove that the factorization of an integer greater than 1 into prime factors is unique. To do so, we shall build on the properties of the integers that we have already proved. We shall also need some other results.

We state the main result now, to help you see why we develop the tools that we do.

> **Theorem 5.1 Unique prime factorization**
>
> Let n be an integer greater than 1. Then n has a unique factorization of the form
> $$n = p_1^{k_1} p_2^{k_2} \ldots p_r^{k_r}, \quad p_1 < p_2 < \cdots < p_r,$$
> where $p_1, \ldots, p_r$ are (distinct) primes and $k_1, \ldots, k_r$ are positive integers.
>
> The unique factorization is known as the **prime decomposition** of n.

In Section 4, in Theorem 4.3, we showed that, under the hypothesis of Theorem 5.1, n could be written as some product of primes. Collecting like terms and arranging the primes in ascending order produces an expression of the required form. What we need to do now is to show that any two such expressions for n are actually the same.

The basis of our argument is the following.

> **Lemma 5.1**
>
> Let p be a prime, and let a and b be integers such that
> $$p \mid ab.$$
> Then
> $$p \mid a \quad \text{or} \quad p \mid b.$$

Proof

Assume that p does not divide a. We shall show that, under these circumstances, p must divide b.

Since p does not divide a, we know, from Exercise 4.4, that
$$\text{hcf}\{p, a\} = 1.$$
From Theorem 2.1 there exist integers x and y such that
$$1 = xp + ya.$$
Multiplying by b gives
$$b = xpb + yab$$
$$= (xb)p + y(ab).$$
So b is an integer combination of p and ab, both of which are divisible by p. Hence, by Exercise 1.6, p divides b. ∎

The following exercises ask you to generalize Lemma 5.1 in two ways.

Exercise 5.1

Let p be a prime and $a_1, \ldots, a_n$ be integers such that
$$p \mid a_1 \ldots a_n.$$
Prove that p divides one of $a_1, \ldots, a_n$.

Hint Use the Principle of Mathematical Induction.

Exercise 5.2

Let a, b and n be integers such that
$$n \mid ab \quad \text{and} \quad \text{hcf}\{a, n\} = 1.$$
Show that
$$n \mid b.$$

The next exercise is a deduction from the definition of prime, and provides the final tool that we need to prove Theorem 5.1.

Exercise 5.3

Let p and q be primes such that p divides q. Prove that $p = q$.

We now use our various results, together with the minimal counterexample strategy, to prove Theorem 5.1.

Proof of Theorem 5.1

Suppose that n is the smallest integer greater than 1 which has two distinct factorizations into primes of the prescribed form. That is,
$$n = p_1^{k_1} \ldots p_r^{k_r} = q_1^{l_1} \ldots q_s^{l_s},$$
where
$$p_1 < \cdots < p_r \quad \text{and} \quad q_1 < \cdots < q_s$$
are primes and
$$k_1, \ldots, k_r, l_1, \ldots, l_s$$
are positive integers.

Since
$$n = p_1^{k_1} \ldots p_r^{k_r} = p_1(p_1^{k_1 - 1} \ldots p_r^{k_r})$$
we have
$$p_1 \mid n.$$
Because $n = q_1^{l_1} \ldots q_s^{l_s}$, it follows that p_1 divides
$$q_1^{l_1} \ldots q_s^{l_s} = \underbrace{q_1 \ldots q_1}_{l_1 \text{ terms}} \underbrace{q_2 \ldots q_2}_{l_2 \text{ terms}} \ldots \underbrace{q_s \ldots q_s}_{l_s \text{ terms}}.$$

By Exercise 5.1, it follows that p_1 divides one of the q_js. Therefore, by Exercise 5.3, it is equal to one of the q_js. Since q_1 is the smallest of the q_js, we have
$$q_1 \leq p_1.$$
The corresponding argument with the ps and qs interchanged gives
$$p_1 \leq q_1.$$
So $p_1 = q_1$.

Hence dividing n by p_1 ($= q_1$) gives
$$n_1 = p_1^{k_1 - 1} \ldots p_r^{k_r} = q_1^{l_1 - 1} \ldots q_s^{l_s}.$$
There are two possibilities: $n_1 = 1$ and $n_1 > 1$.

In the first case
$$n = p_1 = q_1,$$
which contradicts the assumption that the two factorizations were different.

In the second case, since n is a counterexample, its factorizations are different, and hence so are the factorizations of n_1 that we have obtained. But the fact that $n_1 < n$ contradicts the minimality of n, and this completes the proof. ∎

If either the exponent of p_1 or of q_1 is now zero, then remove that term so that the expressions are of the required form, i.e. with all exponents as positive integers.

The following final set of exercises ask you to prove some results that follow from the Unique Prime Factorization Theorem and also to investigate how HCFs and LCMs are related to prime decompositions.

Exercise 5.4

Let p be a prime and k and l be non-negative integers. Prove that
$$p^k \mid p^l \iff k \leq l.$$

Exercise 5.5

Let a and b be positive integers with prime decompositions
$$a = p_1^{k_1} \ldots p_r^{k_r}$$
and
$$b = q_1^{l_1} \ldots q_s^{l_s}.$$
Show that a divides b if, and only if, every prime in the decomposition of a appears in the decomposition of b and its exponent in a is less than or equal to its exponent in b.

Exercise 5.6

Show that, if two integers a and b have a common factor greater than 1, then they have a common prime factor.

Exercise 5.7

Let p be prime and a be a non-negative integer. Suppose that p^n is the highest power of p which divides a. Show that
$$a = p^n b,$$
where p and b are coprime.

Hint Use a proof by contradiction.

Exercise 5.8

Using the same notation as in Exercise 5.5, show that the HCF of positive integers a and b consists of the product of the primes common to both decompositions, with exponents equal to the minimum of the corresponding exponents in the two decompositions.

SOLUTIONS TO THE EXERCISES

Solution 1.1

S is bounded above, by the integer M say, so that

$$s \leq M, \quad \text{for all } s \in S.$$

Hence

$$0 \leq M - s, \quad \text{for all } s \in S,$$

and thus

$$T = \{M - s : \text{for all } s \in S\}$$

consists of non-negative integers.
Therefore, by the extended version of Theorem 1.2, T has a least element, l_T say.

We know that $l_T \in T$. Hence

$$l_T = M - g_S, \quad \text{for some } g_S \in S.$$

Furthermore,

$$l_T \leq M - s, \quad \text{for all } s \in S.$$

Therefore

$$M - g_S \leq M - s, \quad \text{for all } s \in S,$$

giving

$$-g_S \leq -s, \quad \text{for all } s \in S,$$

and hence

$$s \leq g_S, \quad \text{for all } s \in S.$$

Hence $g_S \in S$ and $s \leq g_S$ for all $s \in S$.

Thus g_S is the greatest element of S.

An alternative proof, that makes use of part (a) of the theorem, is the following.

Since S is bounded above, there exists an integer M such that $s \leq M$ for all $s \in S$. Hence the set $-S$ defined by

$$-S = \{-s : s \in S\}$$

is bounded below by $-M$. Therefore, by part (a) of Theorem 1.3, $-S$ has a least element, $-g$ say, where

$$-g \leq -s, \quad \text{for all } s \in S.$$

Therefore $g \in S$ (by definition of $-S$) and

$$s \leq g, \quad \text{for all } s \in S,$$

and hence g is the greatest element of S.

Solution 1.2

Using the first part of the hint, there exist integers q and r such that
$$|a| = bq + r, \quad 0 \le r < b.$$
Since $|a| = -a$, this gives
$$\begin{aligned}a &= -|a| \\ &= -(bq + r) \\ &= b(-q) + (-r).\end{aligned}$$
This is of the correct form, except that $-r$ may be negative.

If $r = 0$ then we can take
$$\tilde{q} = (-q), \quad \tilde{r} = r = 0,$$
and we have
$$a = b\tilde{q} + \tilde{r}, \quad 0 \le \tilde{r} < b.$$
We are left with the case $0 < r < b$.
Using the second part of the hint, we have
$$0 < b - r < b,$$
which suggests that we rearrange the expression for a above as
$$\begin{aligned}a &= b(-q) + (-r) \\ &= b(-q) - b + b - r \\ &= b(-1 - q) + (b - r).\end{aligned}$$
Now, if we take
$$\tilde{q} = (-1 - q) \quad \text{and} \quad \tilde{r} = (b - r),$$
this gives the required result.

Solution 1.3

Using the hint, we build on the result for positive values of b by considering
$$|b| = -b > 0.$$
By Theorem 1.1, there exist integers $\tilde{q}$ and $\tilde{r}$ such that
$$a = |b|\tilde{q} + \tilde{r}, \quad 0 \le \tilde{r} < |b|.$$
Since $|b| = -b$, we have
$$\begin{aligned}a &= (-b)\tilde{q} + \tilde{r} \\ &= b(-\tilde{q}) + \tilde{r}.\end{aligned}$$
If we define
$$q = -\tilde{q} \quad \text{and} \quad r = \tilde{r},$$
then we have found integers satisfying the requirements.

The proof of the uniqueness of q and r is identical to the proof in Theorem 1.1 except for the replacement of b by $|b|$.

Solution 1.4

If zero divides the integer a then there exists an integer q such that
$$a = 0 \times q = 0.$$
Thus $a = 0$, as required.

Solution 1.5

Because x divides y, there exists an integer q_1 such that
$$y = xq_1.$$
Because y divides z, there exists an integer q_2 such that
$$z = yq_2.$$
Combining these, we have
$$z = x(q_1 q_2).$$
As $q_1 q_2 \in \mathbb{Z}$, it follows that x divides z.

Solution 1.6

Applying the definition twice, there exist integers q_1 and q_2 such that
$$a = cq_1 \quad \text{and} \quad b = cq_2.$$
Hence
$$\begin{aligned}\alpha a + \beta b &= \alpha(cq_1) + \beta(cq_2) \\ &= c(\alpha q_1 + \beta q_2).\end{aligned}$$
As $(\alpha q_1 + \beta q_2) \in \mathbb{Z}$, this completes the proof.

Solution 2.1

We first note that $|a| = \pm a$. We shall use this fact to show that a and $|a|$ have exactly the same set of factors.

Suppose that u divides a. Then $a = uq$, for some integer q, and so
$$|a| = \pm uq = u(\pm q).$$
Hence u divides $|a|$. So every factor of a is a factor of $|a|$.

Similarly, from the fact that $a = \pm|a|$, every factor of $|a|$ is a factor of a.

So the integers a and $|a|$ have the same set of factors.

In the same way, b and $|b|$ have the same set of factors.

It follows that, for example, the common factors of a and b are the same as the common factors of a and $|b|$, and so
$$\text{hcf}\{a,b\} = \text{hcf}\{a,|b|\}.$$
The other results follow in the same way.

Solution 2.2

If

We are given integers α and β such that
$$1 = \alpha a + \beta b.$$
Since 1 is the smallest positive integer, and 1 is an integer combination of a and b, it is the smallest, positive, integer combination of a and b. Hence, by Theorem 2.1, $\text{hcf}\{a,b\} = 1$ and a and b are coprime.

Only if

If a and b are coprime then $\text{hcf}\{a,b\} = 1$. By Theorem 2.1, $\text{hcf}\{a,b\}$ is an integer combination of a and b. Thus
$$1 = \alpha a + \beta b,$$
for some integers α and β.

Solution 2.3

Using Theorem 2.1, there exist integers α and β such that
$$d = \alpha a + \beta b.$$
Hence
$$d = \alpha dx + \beta dy,$$
and, dividing by d throughout, gives
$$1 = \alpha x + \beta y.$$
Hence, by the result of the previous exercise, x and y are coprime.

Solution 2.4

Since $\operatorname{hcf}\{a, b\} = 1$ there exist integers α and β such that
$$1 = \alpha a + \beta b.$$
Multiplying both sides by c gives
$$c = c\alpha a + c\beta b$$
$$= (\alpha c)a + \beta(bc).$$
Thus c is an integer combination of a and bc, both of which are divisible by a. Hence, by Exercise 1.6,
$$a \mid c.$$

Solution 2.5

Because $\operatorname{hcf}\{a, b\} = 1$, we can write
$$1 = \alpha a + \beta b.$$
Multiplying throughout by c gives
$$c = \alpha ac + \beta bc.$$
Because a divides c and b divides c, we have
$$c = ax \quad \text{and} \quad c = by,$$
for some integers x and y.
Substituting,
$$c = \alpha a(by) + \beta b(ax)$$
$$= ab(\alpha y + \beta x).$$
Hence ab divides c.

Solution 2.6

The strategy is to show that *some* positive common multiple exists, in which case there must be a *smallest* positive common multiple, by the well-ordering property of $\mathbb{N}$.

Since both a and b are non-zero, it follows that $|ab|$ is positive. Also
$$|ab| = \pm ab = a(\pm b) = b(\pm a),$$
and is, therefore, a multiple of both a and b.

So the set of positive common multiples of a and b is non-empty and hence has a smallest element.

Solution 2.7

As suggested by the hint, we apply the usual strategy of using the Quotient–remainder Theorem and then showing that the remainder is zero.

Since, by definition, $m = \text{lcm}\{a,b\}$ is greater than zero, m is non-zero, and so by Theorem 1.4 we can write
$$n = mq + r, \quad 0 \leq r < |m| = m.$$
Now
$$\begin{aligned}r &= n - mq \\ &= 1 \times n + (-q) \times m,\end{aligned}$$
and so r is an integer combination of m and n. As both a and b divide m and n, r is divisible by a and b (by Exercise 1.6).

Thus r is a non-negative common multiple of a and b which is less than m. Since m is the smallest *positive* common multiple, r cannot be positive.

It follows that $r = 0$ and so m divides n.

Solution 2.8

(a) Since, in $\mathbb{Z}_4$, $2 \neq 0$ and $2 + 2 = 2 \times 2 = 0$, the order of 2 in $\mathbb{Z}_4$ is 2.

Since, in $\mathbb{Z}_6$, $2 \neq 0$, $2 + 2 = 2 \times 2 \neq 0$ and $2 + 2 + 2 = 3 \times 2 = 0$, the order of 2 in $\mathbb{Z}_6$ is 3.

Since, in $\mathbb{Z}_4 \times \mathbb{Z}_6$,
$$\begin{aligned}1 \times (2,2) &= (2,2) \neq (0,0), \\ 2 \times (2,2) &= (0,4) \neq (0,0), \\ 3 \times (2,2) &= (2,0) \neq (0,0), \\ 4 \times (2,2) &= (0,2) \neq (0,0), \\ 5 \times (2,2) &= (2,4) \neq (0,0), \\ 6 \times (2,2) &= (0,0),\end{aligned}$$
the order of $(2,2)$ is 6.

(b) We tabulate the results for all possible (a,b) in the direct product as follows.

(a,b)	Order of a	Order of b	Order of (a,b)
$(0,0)$	1	1	1
$(0,1)$	1	6	6
$(0,2)$	1	3	3
$(0,3)$	1	2	2
$(0,4)$	1	3	3
$(0,5)$	1	6	6
$(1,0)$	4	1	4
$(1,1)$	4	6	12
$(1,2)$	4	3	12
$(1,3)$	4	2	4
$(1,4)$	4	3	12
$(1,5)$	4	6	12
$(2,0)$	2	1	2
$(2,1)$	2	6	6
$(2,2)$	2	3	6
$(2,3)$	2	2	2
$(2,4)$	2	3	6
$(2,5)$	2	6	6
$(3,0)$	4	1	4
$(3,1)$	4	6	12
$(3,2)$	4	3	12
$(3,3)$	4	2	4
$(3,4)$	4	3	12
$(3,5)$	4	6	12

Solution 2.9

To simplify notation we let $l = \text{lcm}\{m, n\}$.

We shall show that $(g, h)^l = (e, e)$ and that no smaller positive power produces the identity.

Note that we use e to represent the identity of both G and H.

We know that
$$l = mx \quad \text{and} \quad l = ny,$$
for some integers x and y. We also know that
$$g^m = e \quad \text{and} \quad h^n = e.$$
Hence
$$\begin{aligned}(g, h)^l &= (g^l, h^l) \\ &= (g^{mx}, h^{ny}) \\ &= ((g^m)^x, (h^n)^y) \\ &= (e^x, e^y) \\ &= (e, e).\end{aligned}$$

Now suppose that $(g, h)^k = (e, e)$, for some positive integer k. It follows that
$$(g, h)^k = (g^k, h^k) = (e, e)$$
and so
$$g^k = e \quad \text{and} \quad h^k = e.$$

Because $g^k = e$ and g has order m, it follows (by Lemma 1.2) that m divides k. Similarly, $h^k = e$ implies that n divides k.

Hence k is a positive common multiple of m and n.
Since l is the least common multiple, it follows that $l \le k$ as required.

Solution 3.3

(a) We set out the calculations as in the boxed version of the Euclidean Algorithm on page 22:
$$\begin{aligned}421 &= 111 \times 3 + 88 \\ 111 &= 88 \times 1 + 23 \\ 88 &= 23 \times 3 + 19 \\ 23 &= 19 \times 1 + 4 \\ 19 &= 4 \times 4 + 3 \\ 4 &= 3 \times 1 + 1 \\ 3 &= 1 \times 3\end{aligned}$$

The last non-zero remainder is 1 and so the HCF is 1.
Thus 111 and 421 are coprime.

(b) We work back from the penultimate line of the Euclidean Algorithm as follows:
$$\begin{aligned}1 &= 4 - 3 \times 1 \\ &= 4 - (19 - 4 \times 4) = 5 \times 4 - 19 \\ &= 5 \times (23 - 19 \times 1) - 19 = 5 \times 23 - 6 \times 19 \\ &= 5 \times 23 - 6 \times (88 - 23 \times 3) = 23 \times 23 - 6 \times 88 \\ &= 23 \times (111 - 88 \times 1) - 6 \times 88 = 23 \times 111 - 29 \times 88 \\ &= 23 \times 111 - 29 \times (421 - 111 \times 3) = 110 \times 111 - 29 \times 421\end{aligned}$$

Hence
$$1 = \text{hcf}\{111, 421\} = 110 \times 111 + (-29) \times 421.$$

Solution 3.4

Since hcf$\{144, 128\} = 16$, we have, by Theorem 2.2,

$$16 \times \text{lcm}\{144, 128\} = 144 \times 128.$$

Hence

$$\text{lcm}\{144, 128\} = 9 \times 128 = 1152.$$

Solution 4.1

Suppose G is a group of order p. Since any prime is greater than 1, there exists some non-identity element g of G. The cyclic subgroup

$$H = \langle g \rangle$$

generated by g contains the element g and also the identity element e, and therefore has more than one element.

By Lagrange's Theorem, the number of elements in H is a (positive) factor of the number of elements in G, namely p. As p is prime it has only two positive factors, 1 and p. Since H has more than one element, it must have p elements, and therefore

$$H = G.$$

Since H is cyclic, by definition, so is G.

Solution 4.2

Since n is composite, the set of all positive factors of n other than n and 1 is a non-empty set of positive integers. Hence this set has a least element, p say, such that

$$1 < p < n.$$

We now show that p is prime, using contradiction.

Suppose that p is not prime. Then, since $p > 1$ it has a factor, p_1, such that

$$1 < p_1 < p.$$

However, p_1 divides p, which, in turn, divides n. Hence

$$p_1 \mid n, \quad 1 < p_1 < p.$$

This contradicts the choice of p as the smallest such factor of n.

This contradiction completes the proof.

Solution 4.3

The smallest prime is 2 and this is a factor of 1728:

$$1728 = 2 \times 864.$$

Continuing:

$$\begin{aligned}
1728 &= 2 \times 864 \\
&= 2 \times 2 \times 432 \\
&= 2 \times 2 \times 2 \times 216 \\
&= 2 \times 2 \times 2 \times 2 \times 108 \\
&= 2 \times 2 \times 2 \times 2 \times 2 \times 54 \\
&= 2 \times 2 \times 2 \times 2 \times 2 \times 2 \times 27 \\
&= 2 \times 2 \times 2 \times 2 \times 2 \times 2 \times 3 \times 9 \\
&= 2 \times 2 \times 2 \times 2 \times 2 \times 2 \times 3 \times 3 \times 3.
\end{aligned}$$

We would normally write this result as
$$1728 = 2^6 3^3.$$

Solution 4.4

The only positive factors of p are p and 1. Since p does not divide a, the *only* positive common factor that p and a can have is 1. Thus hcf$\{p, a\} = 1$, that is, they are coprime.

Solution 4.5

The only positive factors of q are 1 and q. As p is neither 1 nor q, it does not divide q. By applying the result of the previous exercise with $a = q$, we have that p and q are coprime.

Solution 4.6

No, a and c need not be coprime. A suitable counterexample is given by
$$a = 2, \quad b = 3, \quad c = 4.$$

Solution 5.1

Using the hint, we proceed using the Principle of Mathematical Induction on the number of terms, n, in the product.

If $n = 1$, that is p divides a_1, then the result is trivially true.

Now assume the result for $n = k$ where $k \geq 1$, and suppose that
$$p \mid a_1 \ldots a_{k+1}.$$

We may rewrite the product as
$$(a_1 \ldots a_k) a_{k+1},$$
in other words, as the product of two integers.

Applying Lemma 5.1, p must divide one of
$$(a_1 \ldots a_k) \quad \text{or} \quad a_{k+1}.$$

If p divides the first of these then, by the induction hypothesis, p divides one of $a_1, \ldots, a_k$ and hence one of $a_1, \ldots, a_{k+1}$.
On the other hand if p divides a_{k+1} again it divides one of $a_1, \ldots, a_{k+1}$.

This completes the inductive step and the proof.

Solution 5.2

Since $\mathrm{hcf}\{a, n\} = 1$, by Theorem 2.1 there exist integers x and y such that
$$1 = xa + yn.$$

Multiplying by b gives,
$$b = xab + ynb$$
$$= x(ab) + (yb)n.$$

This shows that b is an integer combination of the integers ab and n, both of which are divisible by n. Hence, by Exercise 1.6, n divides b.

Note that, since $\mathrm{hcf}\{a, n\}$ is defined, at least one of a and n must be non-zero, and so we can apply Theorem 2.1.

Solution 5.3

Since p is prime and p divides q, p is a positive factor of q. As q is prime, its only positive factors are 1 and q. As p is prime, it is a positive integer and is not 1. Hence $p = q$.

Solution 5.4

If

If $k \leq l$, then $l - k$ is a non-negative integer. Thus p^k and p^{l-k} are integers such that
$$p^l = p^k p^{l-k},$$
which shows that p^k divides p^l.

Only if

Suppose that p^k divides p^l. Then there exists a positive integer b such that
$$p^l = p^k b.$$
By the uniqueness of prime decomposition, b must either be 1, in which case $l = k$, or have a prime decomposition consisting entirely of ps. In this case
$$b = p^m,$$
for some positive integer m. Hence
$$p^{l-k} = p^m.$$
It follows from the uniqueness of prime decomposition that
$$l - k = m > 0$$
and so $l > k$.

Combining the cases, we have $k \leq l$.

Solution 5.5

If

If every prime in the decomposition of a appears in that for b, with exponent in a not exceeding that in b, then we have
$$a = p_1^{k_1} \ldots p_r^{k_r},$$
$$b = p_1^{l_1} \ldots p_r^{l_r} \ldots p_s^{l_s},$$
where
$$k_1 \leq l_1, \ldots, k_r \leq l_r, \quad r \leq s.$$
Hence
$$d = p_1^{l_1 - k_1} \ldots p_r^{l_r - k_r} \ldots p_s^{k_s}$$
is an integer and
$$b = ad.$$
Hence a divides b.

Only if

Suppose that a divides b and that
$$a = p_1^{k_1} \ldots p_r^{k_r}$$
is the prime decomposition of a.

Then for each $i = 1, \ldots, r$, $p_i^{k_i}$ divides a and hence, by Exercise 1.5, divides b. Thus
$$b = p_i^{k_i} c,$$
for some integer c.

The prime decomposition of b is given by the product
$$p_i^{k_i}(\text{prime decomposition of } c),$$
rearranged as necessary.

Hence p_i must occur in the prime decomposition of b with an exponent at least as big as k_i.

This completes the proof.

Solution 5.6

Suppose that integers a and b have a common factor $d > 1$. Since $d > 1$, it has a prime decomposition. If p is some prime in this decomposition, we have

$$p \mid d, \quad d \mid a \quad \text{and} \quad d \mid b.$$

It follows, from Exercise 1.5, that

$$p \mid a \quad \text{and} \quad p \mid b.$$

Solution 5.7

We prove this by contradiction. Suppose that

$$\text{hcf}\{p, b\} > 1.$$

Then the HCF must be p and so p divides b. We can then write

$$b = p b_1,$$

for some integer b_1. But then

$$a = p^{n+1} b_1$$

and this contradicts the assumption about a.
Hence we must have $\text{hcf}\{p, b\} = 1$, i.e. p and b must be coprime.

Solution 5.8

Suppose that d is constructed as in the question, i.e. d is the product of the primes common to the decompositions of both a and b, with exponents equal to the minimum of the corresponding exponents in the two decompositions. Since all primes in the decomposition of d appear in the decompositions of both a and b, with exponents less than or equal to the exponents in a and b, it follows from Exercise 5.5 that

$$d \mid a \quad \text{and} \quad d \mid b.$$

Thus d is a positive common factor of a and b.

On the other hand, if c is a positive common factor of a and b, then, by Exercise 5.5, any prime p in the decomposition of c occurs in the decompositions of both a and b and, hence, in the decomposition of d.

The exponent of p in c is less than or equal to its exponent in both a and b, and hence is less than or equal to the minimum of these exponents. This minimum is precisely the exponent of p in d.

This argument holds for all primes in the decomposition of c and hence c divides d.

Thus d is a common factor of a and b, divisible by all such common factors. Thus

See result (b) after the proof of Theorem 2.1.

$$d = \text{hcf}\{a, b\}.$$

OBJECTIVES

After you have studied this unit, you should be able to:
(a) apply the well-ordering property of $\mathbb{N}$ to simple proofs;
(b) apply the Quotient–remainder Theorem to simple proofs concerning divisibility properties of integers;
(c) apply the Euclidean Algorithm to find HCFs and LCMs;
(d) find and apply the integer combination form of HCFs;
(e) apply the definition and divisibility properties of primes to simple proofs.

INDEX

bounded 6
bounded above 6
bounded below 6
common divisor 17
common factor 11
common multiple 14
composite 19
coprime 14
divides 5, 9
divisor 5, 9
Euclidean algorithm 17

factor 5, 9
greatest common divisor 11
HCF 11
hcf$\{a, b\}$ 11
highest common factor 11
integer combination 10
LCM 14
lcm$\{a, b\}$ 14
lower bound 6
lowest common multiple 14
minimal counterexample strategy 19

multiple 9
prime 18
prime decomposition 21
prime factor 19
quotient 5
quotient–remainder theorem 5, 8
remainder 5
unique prime factorization theorem 21
upper bound 6
well-ordering property of $\mathbb{N}$ 5